Does "Excess" Dietary Salt Cause Cardiovascular Toxicity?

WRITTEN BY

Lila Abassi, M.D.

A publication of the

AMERICAN COUNCIL ON SCIENCE AND HEALTH

Publisher name: American Council on Science and Health
Title: Does "Excess" Dietary Salt Cause Cardiovascular Toxicity?
Author: Lila Abassi, M.D.
Subject (general): Science and Health
Publication Year: 2016
Binding Type (i.e. perfect (soft) or hardcover): Perfect
ISBN: 978-0-9910055-5-0

Cover image credit to Shutterstock.

Acknowledgements

The American Council on Science and Health appreciates the contributions of the reviewers named below:

Josh Bloom Ph.D.
Director of Chemical and Pharmaceutical Sciences
American Council on Science and Health

Hank Campbell
President
American Council on Science and Health

Idrees Danishpajooh, M.D.
Memorial Hospital Jacksonville

Magdalene Fiddler, M.D.
Avera McKennan Hospital and University Health Center

Jason Gritti, M.D.
Kaiser Perminente South Sacramento

Robert B. Sklaroff, M.D., F.A.C.P.
Philadelphia, PA

Table of Contents

Public health mandates are, too often, based upon shaky scientific data. Although the population-sample with hypertension may be treated by diuretics—drugs that remove salt—epidemiologic studies have yet to show that the health of the general population is harmed by unlimited salt-intake. Challenging this shibboleth[1] illustrates how ideological politicians maul consensus viewpoints.

Indeed, the American Heart Association and the National Institutes of Health advise restricting daily intake to less than 2300 mg. of salt (the equivalent of one teaspoon of table salt), and to less than 1500 mg. per day in patients with conditions such as hypertension, congestive heart failure, cirrhosis and kidney disease[2]. (Sodium chloride—a.k.a. table salt—is 40% sodium.) The average American diet contains about 3400 mg. of sodium per day.

They cannot, however, identify epidemiologic support for this recommendation and, indeed, profound restriction of dietary salt may be associated with adverse health outcomes. In fact, although such sweeping proclamations may appear reasonable, they ignore the tremendous heterogeneity in the way each individual responds to ingesting salt.

1   Taubes, G. (1998). The (Political) Science of Salt. Retrieved from https://www.stat.
    berkeley.edu/~rice/Stat2/salt.html

2   About Sodium (Salt). (2015). Retrieved from http://www.heart.org/HEARTORG/
    GettingHealthy/NutritionCenter/HealthyEating/About-Sodium-Salt_UCM_463416_Article.
    jsp#.Vi_UCrRVhBc

1
The role of salt in the human body

To maintain sufficient tissue-perfusion, humans stabilize blood volume (and salt concentration) by retaining water if confronted with salt-overload and by excreting water if deprived of salt. Complex hormonal and neuroendocrine cascades mediate this process, which may be blunted by disease states (genetic or acquired) that predispose to hypertension.

Blood pressure may be elevated by increased circulating-volume (water) or by narrowed arteries (atherosclerosis). Paradoxically, low salt-intake may prompt the kidneys to secrete renin in an effort to retain sodium; this, in turn, may enhance heart toxicity in an otherwise hypertensive patient[3].

3   Al-Haj, Z. (2011). Acid-Base and Fluid Balance. Retrieved from http://slideshare.net/ Zahra_Haj_Issa/acid-base-balance-fluid-balance

Blood pressure, sodium, and water

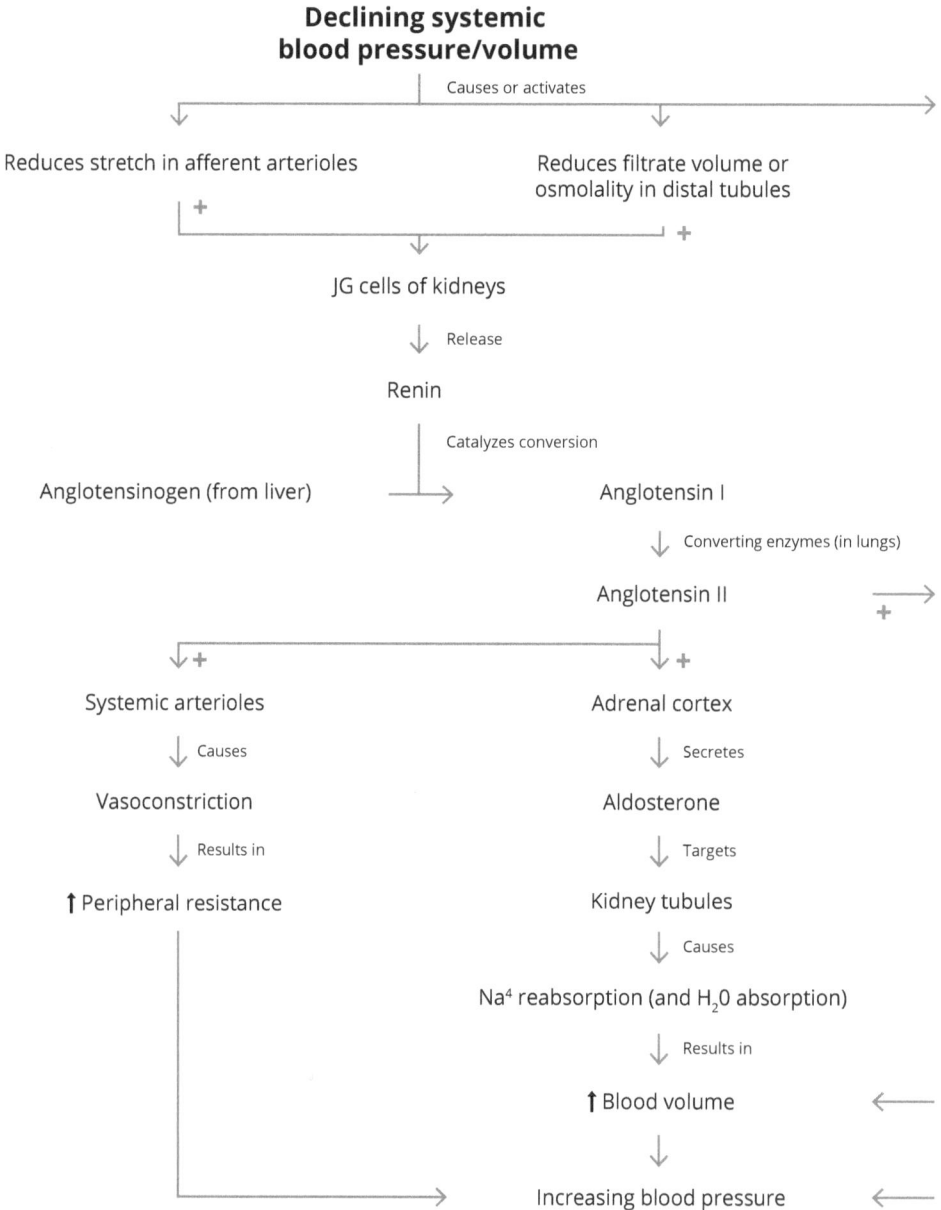

**Declining systemic
blood pressure/volume**

Causes or activates

Reduces stretch in afferent arterioles

Reduces filtrate volume or
osmolality in distal tubules

+   +

JG cells of kidneys

Release

Renin

Catalyzes conversion

Anglotensinogen (from liver)   Anglotensin I

Converting enzymes (in lungs)

Anglotensin II

+

Systemic arterioles	Adrenal cortex
Causes	Secretes
Vasoconstriction	Aldosterone
Results in	Targets
↑ Peripheral resistance	Kidney tubules
	Causes
	Na^4 reabsorption (and H_2O absorption)
	Results in
	↑ Blood volume
	↓
	Increasing blood pressure

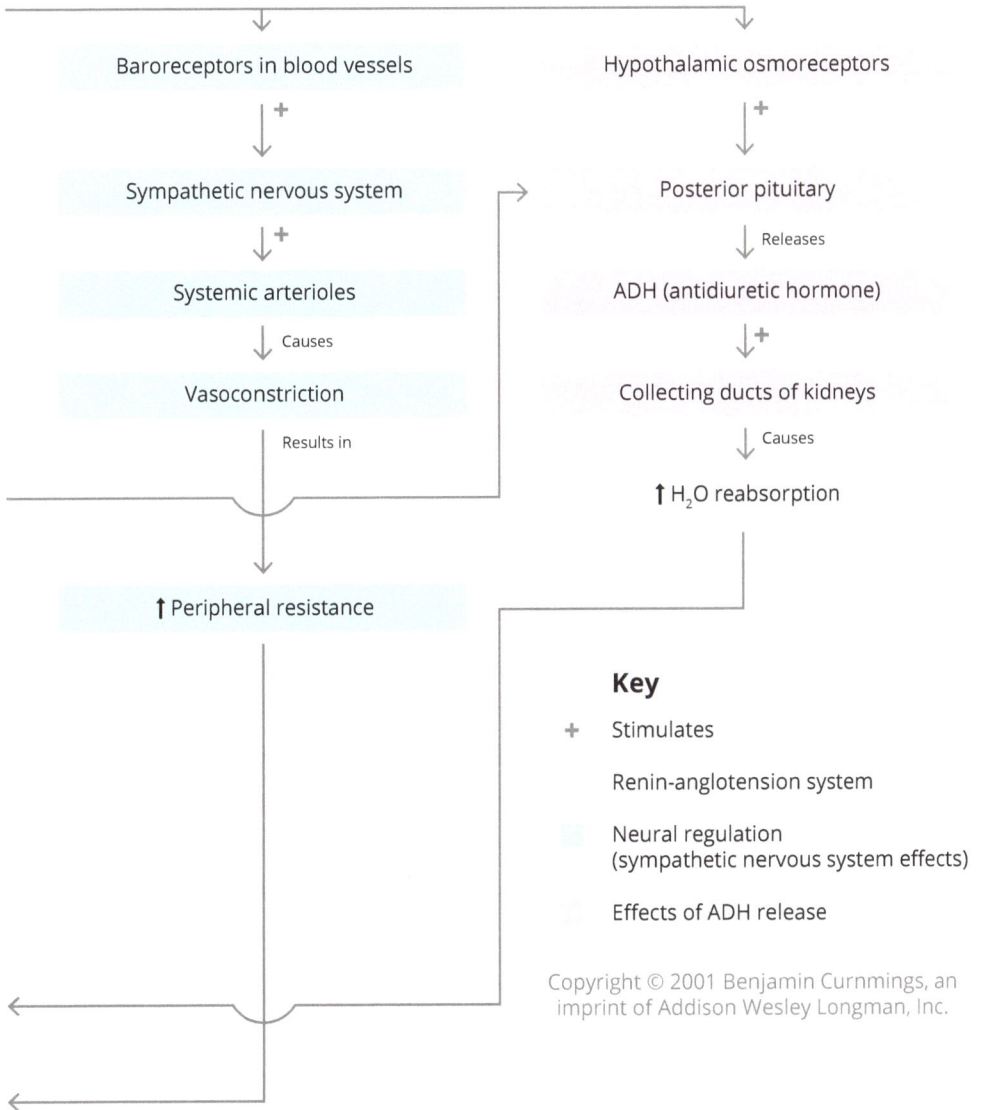

Baroreceptors in blood vessels

+

Sympathetic nervous system

+

Systemic arterioles

Causes

Vasoconstriction

Results in

Hypothalamic osmoreceptors

+

Posterior pituitary

Releases

ADH (antidiuretic hormone)

+

Collecting ducts of kidneys

Causes

↑ H_2O reabsorption

↑ Peripheral resistance

Key

+   Stimulates

Renin-anglotension system

Neural regulation
(sympathetic nervous system effects)

Effects of ADH release

2

Can hypertension – the 'silent killer' – be prevented?

"Dietary Guidelines for Americans, 2010," issued by the Centers for Disease Control and Prevention [CDC][4], recommends that "everyone age 2 and up should consume less than 2300 milligrams (mg.) of sodium each day. Some groups of people should further limit sodium intake to 1500 mg. per day, including:

▶ Adults age 51 or older

▶ All African-Americans

▶ Anyone who has high blood pressure, diabetes, or chronic kidney disease."

It has not been proven that global restriction of dietary-sodium would decrease the incidence of cardiovascular disease and consequent death. Some patients (such as those who retain salt) would benefit from doing so, but others (such as those with rare salt-wasting conditions[5]) would

4   Salt. (2015). Retrieved from http://www.cdc.gov/salt/

5   Alderman, M. (2012). *Einstein on...Salt: Is Less Sodium Always Better?* [Video file]. Retrieved from https://www.youtube.com/watch?v=NbyE40tmSNo

not. For the 5.1 million Americans diagnosed with heart failure, low body salt concentration (hyponatremia) could prove lethal[6].

Hyponatremia associated with reduced survival in patients with severe chronic heart failure

Data from: Lee WH, Packer M. Circulation 1986; 73:257

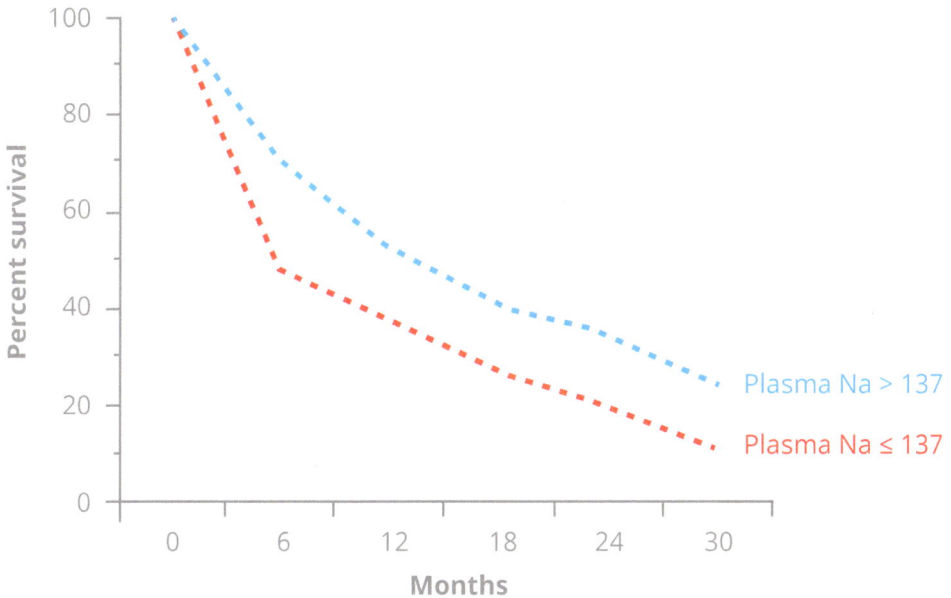

6   Vasan, R. S., & Wilson, P. W. (2015). Epidemiology and causes of heart failure. Retrieved from www.uptodate.com/contents/epidemiology-and-causes-of-heart-failure?source=search_result&search=congestive+heart+failure&selectedTitle+6~150

Survival over time in patients with severe chronic heart failure and a left ventricular ejection fraction less than 30 percent who, at study entry, had either a normal plasma sodium concentration (greater than 137 meq/L, solid line) or hyponatremia (plasma sodium less than or equal to 137 meq/L, dashed line). Survival was significantly reduced in the patients with hyponatremia. The survival rate was very low (approximately 15 percent at 12 months) in those with a baseline plasma sodium concentration less than or equal to 130 meq/L.

This challenge is properly addressed by noting "there is no question that sodium restriction can reduce blood pressure. However, the link between salt restriction and cardiovascular disease has not been adequately established and is heterogeneous at best from evidence from observational studies rather than randomized trials. There may actually be a J-shaped [i.e., a nonlinear] relationship between salt intake and risk, and the contention that salt restriction will reduce cardiovascular risk is an argument of hope over reason"[7]. The greatest effect is seen in individuals who consume more that 5.0 g. per day versus less than 3.0 g. per day, while poor outcomes are observed at extremely low levels.

7   Walker, J., MacKenzie, A. D., & Dunning, J. (2007, August 20, 2007). Does reducing your salt intake make you live longer? [Paper]. *Interactive CardioVascular and Thoracic Surgery*, 6(6), 793-798. http://dx.doi.org/10.1510/icvts.2007.165415

3
Three population-studies provide limited insight

To examine the effect of dietary sodium reduction on cardiovascu-lar events and/or deaths (i.e., angina, myocardial infarction, coronary revascularization, and cerebrovascular accident), sequential "**Trials of Hypertension Prevention**" were conducted: 1987-1990 (TOHP I) and 1990-1995 (TOHP II). In these randomized controlled trials (the gold standard of research), patients were randomized into a sodium interven-tion group and a control group[8].

A reduction of sodium in the diet resulted in a 25-30% lower risk of cardiovascular outcomes during the 10-15 years after the trial, but these short-term benefits were attenuated when assessed in the long-term[9].

Reliance upon these conclusions is problematic because the investi-gators assumed that the patients continued to adhere to mandates in each study-arm. These data were tainted further because follow-up was conducted by survey questionnaires based mostly on self-reported data

8   Cook, N. R., Cutler, J. A., Obarzanek, E., Buring, J. E., Rexrode, K. M., Kumanyika, S. K., ... Wheton, P. K. (2007, April 20). Long term effects of dietary sodium reduction on cardiovascular disease outcomes: observational follow-up of the trials of hypertension prevention (TOHP). *BMJ.* http://dx.doi.org/10.1510/icvts.2007.165415

9   Alderman, M. (2012). *Einstein on...Salt: Is Less Sodium Always Better?* [Video file]. Retrieved from https://www.youtube.com/watch?v=NbyE40tmSNo

rather than direct measurement. People tend to paint a rosier picture of results than what is actually the case ("social desirability" bias).

The "**Prospective Urban Rural Epidemiology**" (PURE) Study included more than 100,000 adults sampled from the general population of 17 countries, measuring daily sodium excretion because it reflected sodium consumption (we excrete what we consume). While 10% had low excretion (< 3.00 g.)—4% of whom having excretion in the range currently recommended by current CDC-guidelines—the remaining 90% had moderate (3.00-5.99 g.) and high (> 5.99 g.) excretion[10].

Patients with the low and high levels of sodium excretion had an increased risk for adverse cardiovascular events, and a positive, albeit non-uniform, relationship detected between sodium excretion and blood pressure:

▸ Strongest in high range,

▸ Modest in moderate range,

▸ Nonsignificant in low range.

The study lacks an intervention arm; in the absence of a control-group, it is not possible to establish cause-and-effect. Because such a small proportion of the world's population consumes sodium in the low range, sodium intake is unrelated to blood pressure in this sample. As a result, it is necessary to question the validity of overall recommendations for reduced sodium intake as a strategy to deal with hypertension. Thus, it cannot be claimed that reducing sodium intake will improve blood pressure and cardiovascular outcomes generally.

---

10 Yusuf, S., Rangarajan, S., Teo, K., Islam, S., Li, W., Bo, J., ... Gupta, R. (2014, August 28). Cardiovascular risk and events in 17 low-, middle-, and high-income countries. *The New England Journal of Medicine*, 371, 818-27. http://dx.doi.org/10.1056/NEJMoa1311890

The "**Global Burden of Diseases Nutrition and Chronic Diseases Expert Group**" (Nutricode) study entailed meta-analysis of 107 randomized studies from 66 countries, surveying the impact on blood pressure of sodium intake, as determined by urinary excretion and diet. The authors found a linear, dose-response inverse relationship (P<0.001) between sodium reduction and reduction in blood pressure. "On the basis of correlations between sodium intake and between blood pressure and cardiovascular mortality," they determined that 1.65 million deaths worldwide in 2010 were attributable to sodium consumption above 2.0 g. per day (95% uncertainty interval, 1.10 to 2.22 million)[11]."

There was no direct link between sodium reduction and mortality, and there was no linear relationship either between sodium and blood pressure or between blood pressure and cardiovascular disease. "The validity of this analysis requires a linear relation between A and B, B and C, and A and C" but there are other pathophysiological considerations[12]. Such major flaws serve to invalidate the data.

11 Mozaffarian, D., Fahimi, S., Singh, G. M., Micha, R., Khatibzadeh, S., Engell, R. E., ... Powles, J. (2014, August 14). Global sodium consumption and death from cardiovascular causes. *The New England Journal of Medicine*, 371, 624-34. http://dx.doi.org/10.1056/NEJMoa1304127

12 Alderman, M. (2014, November 27). Sodium and Cardiovascular Disease [Letter to the editor]. *The New England Journal of Medicine*, 371(22), 2137. http://dx.doi.org/10.1056/NEJMc1412113

4
Two systematic reviews are also inconclusive

After **Danes** adjusted the data for confounders (findings that can cause bias), they found that all-cause mortality was higher when people consumed less (versus usual) sodium. There was a U-shaped relationship between the sodium intake and blood pressure. Their "dose-response" analysis revealed a more favorable health outcome among people with sodium consumption in the range of 2645 – 4945 mg./day, intake that coincides with the amount of sodium consumed by most of the world absent any intervention[13].

The "**Institute of Medicine**" (IOM) assembled an expert panel to review the body of existing literature to assess potential benefits and adverse outcomes of reducing sodium intake on the population[14]. It found:

▸ High sodium intake had a positive relationship with cardiovascular disease, when blood pressure was used as a surrogate marker for CVD and stroke.

13 Graudal, N., Jurgens, G., Baslund, B., & Alderman, M. (2014, January 24). Compared with usual sodium intake, low - and excessive- sodium diets are associated with inreased mortality: A meta-analysis. *American Journal of Hypertension*. http://dx.doi.org/10.1093/ajh/hpu028

14 Sodium intake in populations: Assessment of Evidence. (2013). Retrieved from http://iom.nationalacademies.org/~/media/Files/Report%20Files/2013/%20 Sodium-Intake-Populations/SodiumIntakeinPopulations_RB.pdf

▸ There was insufficient consistency in quality and quantity of data to advocate for daily sodium consumption less than 2300 mg., regarding risk of heart disease, stroke, or all-cause mortality.

▸ Patients with moderate to severe congestive heart failure receiving aggressive medical management suffered greater risk of adverse health effects after ingesting a low-sodium diet.

▸ Patients with diabetes, kidney disease or cardiovascular disease suffered some evidence of poor outcomes—and no evidence of benefit—after ingesting 1500 – 2300 mg. of sodium daily.

▸ There are insufficient data to support recommending that patients in disease subgroups should be treated differently than the general population, such as lowering daily sodium intake to or even below 1500 mg.

5
The politics of salt

In September 2015, New York State and New York City became the first American state and city, respectively, to require restaurants to disclose those items on their menu exceeding the recommendation of 2300 mg. as the daily sodium allowance. This proposal was to take effect on December 1st. (The proposed warning label would be in the form of salt-shaker symbols.) The city Board of Health, proposed by current Mayor Bill DeBlasio, supported this measure unanimously; it followed efforts of former NYC Mayor Michael Bloomberg to acquire voluntary pledges from restaurants[15].

The measure would affect about 10% of menu items at NYC franchises with at least 15 chains nationwide, constituting about 1/3 of all the city's restaurant traffic. Some advocates of public health are applauding this move[16]. "This really represents, to me, the next step in allowing usable information for our community to make better health decisions. My hope is that this impacts not only consumer practices but also impacts the practices of our restaurants."

This measure was criticized by the restaurant industry as being costly and unnecessary, although the New York State Restaurant Association has not decided whether it will challenge it legally[17]. "This is just the latest in a long litany of superfluous hoops that restaurants here in New York must jump through....every one of these cumbersome new laws makes it tougher and tougher for restaurants to find success."

15 Peltz, J. (2015). NYC to require salt warnings on menus. Retrieved from http://www.usatoday.com/story/money/business/2015/09/09/nyc-salt-warnings/71936828/

16 Peltz, J. (2015). NYC to require salt warnings on menus. Retrieved from http://www.usatoday.com/story/money/business/2015/09/09/nyc-salt-warnings/71936828/

17 Goldman, H. (2015). New York City requires chain restaurants to post warnings on salty food. Retrieved from http://www.bloomberg.com/politics/articles/2015-09-09/nyc-requires-chain-restaurants-to-post-warnings-on-salty-food

Conclusion

Experts predictably agree that randomized, controlled trials are needed, to compare diets with normal vs. reduced sodium. One researcher editorialized[18], "In the absence of such a trial, the results argue against the reduction of dietary sodium as an isolated public health recommendation." Another researcher, commenting on the Danish study[19], opined, "For science to advance, from time-to-time, medical textbooks and dogma's need a Copernican revolution."

NYC Board of Health officials want people to "know" what is in their food, rather than depending upon consumers to say "no." Yet, after calorie-counts became menu requirements, people have not chosen to consume fewer calories[20]. And such "educational" measures can lead to efforts to ban mega-sized soda containers, an intervention that was blocked in NYC.

Nevertheless, absent scientific support, politicians advocate such extreme measures and costly mandates to the detriment of businesses. Better they should, for example, enhance screening for hypertension,

18 Oparil, S. (2014, August 14). Low sodium intake - cardiovascular health benefit or risk? [Editorial]. *The New England Journal of Medicine*, 371(7), 677-79. http://dx.doi.org/10.1056/NEJMe1407695

19 Asayama, K., Stolarz-Skrzypek, K., Persu, A., & Staessen, J. (2014, September). Systematic review of health outcomes in relation to salt intake highlights the widening divide between guidelines and the evidence [Commentary]. *American Journal of Hypertension*, 27(9), 1138-42. http://dx.doi.org/10.1093/ajh/hpu126

20 Starapoli, N. (2015). Expect only frowns if silly happy meal law passes. Retrieved from http://acsh.org/2015/08/expect-only-frowns-if-silly-happy-meal-law-passes/

which is undiagnosed in 20% of American adults[21]; this is a diagnosis that is treatable, and chronic management prevents cardiovascular disease.

Science must not be forced to conform to politics; rather, policies must be shaped to reflect evidence-based data accurately.

21 Yoon, S. S., Burt, V., Louis, T., & Carroll, M. D. (2012). Hypertension among adults in the United States, 2009-2010. Retrieved from http://www.cdc.gov/nchs/data/databriefs/db107.htm

References

About Sodium (Salt). (2015). Retrieved from http://www.heart.org/HEARTORG/GettingHealthy/NutritionCenter/HealthyEating/About-Sodium-Salt_UCM_463416_Article.jsp#.Vi_UCrRVhBc

Al-Haj, Z. (2011). Acid-Base and Fluid Balance. Retrieved from http://slideshare.net/Zahra_Haj_Issa/acid-base-balance-fluid-balance

Alderman, M. (2012). *Einstein on...Salt: Is Less Sodium Always Better?* [Video file]. Retrieved from https://www.youtube.com/watch?v=NbyE40tmSNo

Alderman, M. (2012). *Einstein on...Salt: Is Less Sodium Always Better?* [Video file]. Retrieved from https://www.youtube.com/watch?v=NbyE40tmSNo

Alderman, M. (2014, November 27). Sodium and Cardiovascular Disease [Letter to the editor]. *The New England Journal of Medicine*, 371(22), 2137. http://dx.doi.org/10.1056/NEJMc1412113

Asayama, K., Stolarz-Skrzypek, K., Persu, A., & Staessen, J. (2014, September). Systematic review of health outcomes in relation to salt intake highlights the widening divide between guidelines and the evidence [Commentary]. *American Journal of Hypertension*, 27(9), 1138-42. http://dx.doi.org/10.1093/ajh/hpu126

Cook, N. R., Cutler, J. A., Obarzanek, E., Buring, J. E., Rexrode, K. M., Kumanyika, S. K., ... Wheton, P. K. (2007, April 20). Long term effects of dietary sodium reduction on cardiovascular disease outcomes: observational follow-up of the trials of hypertension prevention (TOHP). *BMJ*. http://dx.doi.org/10.1510/icvts.2007.165415

Goldman, H. (2015). New York City requires chain restaurants to post warnings on salty food. Retrieved from http://www.bloomberg.com/politics/articles/2015-09-09/nyc-requires-chain-restaurants-to-post-warnings-on-salty-food

Graudal, N., Jurgens, G., Baslund, B., & Alderman, M. (2014, January 24). Compared with usual sodium intake, low - and excessive- sodium diets are associated with inreased mortality: A meta-analysis. *American Journal of Hypertension*. http://dx.doi.org/10.1093/ajh/hpu028

Mozaffarian, D., Fahimi, S., Singh, G. M., Micha, R., Khatibzadeh, S., Engell, R. E., ... Powles, J. (2014, Auguest 14). Global sodium consumption and death from cardiovascular causes. *The New England Journal of Medicine*, 371, 624-34. http://dx.doi.org/10.1056/NEJMoa1304127

Oparil, S. (2014, August 14). Low sodium intake - cardiovascular health benefit or risk? [Editorial]. *The New England Journal of Medicine*, 371(7), 677-79. http://dx.doi.org/10.1056/NEJMe1407695

Peltz, J. (2015). NYC to require salt warnings on menus. Retrieved from http://www.usatoday.com/story/money/business/2015/09/09/nyc-salt-warnings/71936828/

Salt. (2015). Retrieved from http://www.cdc.gov/salt/

Sodium intake in populations: Assessment of Evidence. (2013). Retrieved from http://iom.nationalacademies.org/~/media/Files/Report%20Files/2013/%20Sodium-Intake-Populations/SodiumIntakeinPopulations_RB.pdf

Starapoli, N. (2015). Expect only frowns if silly happy meal law passes. Retrieved from http://acsh.org/2015/08/expect-only-frowns-if-silly-happy-meal-law-passes/

Taubes, G. (1998). The (Political) Science of Salt. Retrieved from https://www.stat.berkeley.edu/~rice/Stat2/salt.html

Vasan, R. S., & Wilson, P. W. (2015). Epidemiology and causes of heart failure. Retrieved from www.uptodate.com/contents/epidemiology-and-causes-of-heart-failure?source=search_result&search=congestive+heart+failure&selectedTitle+6~150

Walker, J., MacKenzie, A. D., & Dunning, J. (2007, August 20, 2007). Does reducing your salt intake make you live longer? [Paper]. *Interactive CardioVascular and Thoracic Surgery*, 6(6), 793-798. http://dx.doi.org/10.1510/icvts.2007.165415

Yoon, S. S., Burt, V., Louis, T., & Carroll, M. D. (2012). Hypertension among adults in the United States, 2009-2010. Retrieved from http://www.cdc.gov/nchs/data/databriefs/db107.htm

Yusuf, S., Rangarajan, S., Teo, K., Islam, S., Li, W., Bo, J., ... Gupta, R. (2014, August 28). Cardiovascular risk and events in 17 low-, middle-, and high-income countries. *The New England Journal of Medicine*, 371, 818-27. http://dx.doi.org/10.1056/NEJMoa1311890

BOARD OF SCIENTIFIC AND POLICY ADVISORS

BOARD OF SCIENTIFIC AND POLICY ADVISORS
(CONTINUED)

BOARD OF SCIENTIFIC AND POLICY ADVISORS
(CONTINUED)

www.ingramcontent.com/pod-product-compliance
Lightning Source LLC
Chambersburg PA
CBHW041719200326
41520CB00001B/160